¡Criaturas diminutas!/Bugs, Bugs, Bugs!

Mariquitas/Ladybugs

por/by Margaret Hall

Traducción/Translation: Dr. Martín Luis Guzmán Ferrer
Editor Consultor/Consulting Editor: Dra. Gail Saunders-Smith

Consultor/Consultant: Gary A. Dunn, MS, Director of Education
Young Entomologists' Society Inc.
Lansing, Michigan

Capstone
press

Mankato, Minnesota

Pebble Plus is published by Capstone Press,
151 Good Counsel Drive, P.O. Box 669, Mankato, Minnesota 56002.
www.capstonepress.com

1 2 3 4 5 6 11 10 09 08 07 06

Library of Congress Cataloging-in-Publication Data
Hall, Margaret, 1947–
    [Ladybugs. Spanish & English]
    Mariquitas = Ladybugs/de/by Margaret Hall.
    p. cm.—(Pebble plus. ¡Criaturas diminutas!/Bugs, bugs, bugs!)
    Includes index.
    ISBN-13: 978-0-7368-6680-4 (hardcover)
    ISBN-10: 0-7368-6680-9 (hardcover)
    1. Ladybugs—Juvenile literature. I. Title. II. Title: Ladybugs. III. Series: Pebble plus.
¡Criaturas diminutas! (Spanish & English)
QL596.C65H34 2007
595.76'9—dc22   2005037469

Summary: Simple text and photographs describe the physical characteristics and habits of ladybugs—in both
    English and Spanish.

Editorial Credits
Sarah L. Schuette, editor; Katy Kudela, bilingual editor; Eida del Risco, Spanish copy editor; Linda Clavel,
    set designer; Kelly Garvin, photo researcher; Karen Hieb, product planning editor

Photo Credits
Bruce Coleman Inc./Gail M. Shumway, 6–7, 20–21; Kim Taylor, 12–13; Raymond Tercafs, 11
Digital Vision, 1
Robert & Linda Mitchell, cover, 5, 8–9, 15
Robert McCaw, 18–19
Stephen McDaniel, 16–17

Note to Parents and Teachers

The ¡Criaturas diminutas!/Bugs, Bugs, Bugs! set supports national science standards related to the diversity of life and heredity. This book describes ladybugs in both English and Spanish. The images support early readers in understanding the text. The repetition of words and phrases helps early readers learn new words. This book also introduces early readers to subject-specific vocabulary words, which are defined in the Glossary section. Early readers may need assistance to read some words and to use the Table of Contents, Glossary, Internet Sites, and Index sections of the book.

Table of Contents

Ladybugs 4

How Ladybugs Look 6

What Ladybugs Do 16

Glossary 22

Internet Sites 24

Index 24

Tabla de contenidos

Las mariquitas 4

Cómo son las mariquitas 6

Qué hacen las mariquitas 16

Glosario 23

Sitios de Internet 24

Índice 24

Ladybugs

What are ladybugs?

Ladybugs are insects

with spots.

Las mariquitas

¿Qué son las mariquitas?

Las mariquitas son unos

insectos que tienen manchitas.

How Ladybugs Look

Most ladybugs are red
or orange.

Cómo son las mariquitas

La mayoría de las mariquitas
son rojas o anaranjadas.

Ladybugs are about the size
of a small pea. Ladybugs
have six legs.

Las mariquitas son como
del tamaño de un chícharo.
Las mariquitas tienen seis patas.

Ladybugs have two antennas. They touch and taste with their antennas.

Las mariquitas tienen dos antenas. Con las antenas tocan y saborean las cosas.

Ladybugs have wings.
Thin wings help ladybugs
fly. Hard wings cover
the thin wings.

Las mariquitas tienen alas.
Las alas delgadas las ayudan a volar.
Las alas duras cubren las alas delgadas.

Ladybugs have sharp jaws.
They bite and chew with
their jaws.

Las mariquitas tienen mandíbulas
filosas. Las mariquitas muerden
y mastican con las mandíbulas.

What Ladybugs Do

Ladybugs sit in the sun.

The sun keeps them warm.

Qué hacen las mariquitas

Las mariquitas se ponen
al sol. El sol hace que
estén calentitas.

Ladybugs lay eggs on plants. Young ladybugs hatch after a few weeks.

Las mariquitas pequeñas eclosionan en unas cuantas semanas.

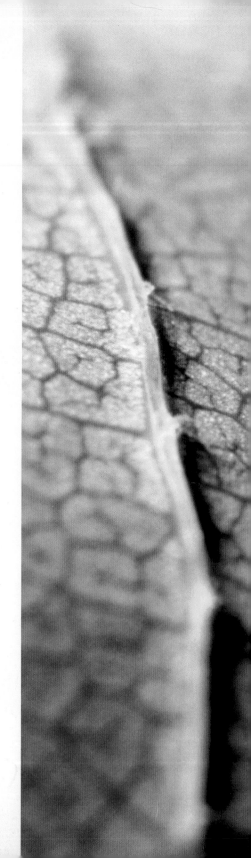

Ladybugs eat aphids.
One ladybug can eat
thousands of aphids
during its lifetime.

Las mariquitas comen
pulgones. Una mariquita
puede comerse miles de
pulgones durante su vida.

Glossary

antenna—a feeler; insects use antennas to sense movement, to smell, and to listen to each other.

aphid—a tiny insect that sucks and eats the juice out of plants

hatch—to break out of an egg

insect—a small animal with a hard outer shell, six legs, three body sections, and two antennas; most insects have wings.

Glosario

la antena—parte del cuerpo para sentir; los insectos usan las antenas para sentir el movimiento, olfatear y escucharse entre sí.

eclosionar—romper el cascarón una larva

el insecto—animal pequeño con un caparazón duro, seis patas, cuerpo dividido en tres secciones y dos antenas; la mayoría de los insectos tiene alas.

el pulgón—insecto pequeñito que chupa y come el jugo de las plantas

Internet Sites

FactHound offers a safe, fun way to find Internet sites related to this book. All of the sites on FactHound have been researched by our staff.

Here's how:

1. Visit *www.facthound.com*

2. Choose your grade level.

3. Type in this book ID **0736866809** for age-appropriate sites. You may also browse subjects by clicking on letters, or by clicking on pictures and words.

4. Click on the **Fetch It** button.

FactHound will fetch the best sites for you!

Index

antennas, 10

aphids, 20

eat, 20

fly, 12

hatch, 18

insects, 4

jaws, 14

legs, 8

lifetime, 20

orange, 6

red, 6

size, 8

spots, 4

sun, 16

taste, 10

touch, 10

wings, 12

Sitios de Internet

FactHound proporciona una manera divertida y segura de encontrar sitios de Internet relacionados con este libro. Nuestro personal ha investigado todos los sitios de FactHound. Es posible que los sitios no estén en español.

Se hace así:

1. Visita *www.facthound.com*

2. Elige tu grado escolar.

3. Introduce este código especial **0736866809** para ver sitios apropiados según tu edad, o usa una palabra relacionada con este libro para hacer una búsqueda general.

4. Haz clic en el botón **Fetch It**.

¡FactHound buscará los mejores sitios para ti!

Índice

alas, 12

anaranjadas, 6

antenas, 10

comer, 20

eclosionan, 18

insectos, 4

manchitas, 4

mandíbulas, 14

patas, 8

pulgones, 20

rojas, 6

saborean, 10

sol, 16

tamaño, 8

tocan, 10

vida, 20

volar, 12